I am a Tyrannosaurus Rex!

Dedication

To the many children (and adults!) who love dinosaurs and use them as a starting point to learn about our Earth's ancient past. Our planet and universe are more marvelous than our minds can conceive. Never stop uncovering Mother Nature's secrets!

Acknowledgements

Thanks to Carol Pouliot, my dear wife of many years, who is a constant source of support and great advice. Thanks also to my sons, John and James, who heard the first inklings of this book (in song) when they were little, and have been my great joys (and gentle critics) ever since. Thanks, too, to the many scientists whose painstaking work—in the field and in the lab—informed my understanding of the strange and wondrous beasts that once roamed the planet. Stephen J. Gould, Robert Bakker and Steve Brusatte, I salute you!

I am a Tyrannosaurus Rex! text and illustrations copyright © 2019 by Jean Edouard Pouliot. All rights reserved.

I am a Tyrannosaurus Rex!

and other poems about dinosaurs

Written and illustrated by
Jean Pouliot

I am a Tyrannosaurus Rex! What I do is not all that complex.
All I do is eat and eat and eat and eat and eat and eat.
I am a Tyrannosaurus Rex!

Stegosaurus, don't sit on my tail!
Or you'll leave a long
 and bloody trail!
Got a walnut for a brain,
Yet kind and gentle, in the main.
Stegosaurus, don't sit on my tail!

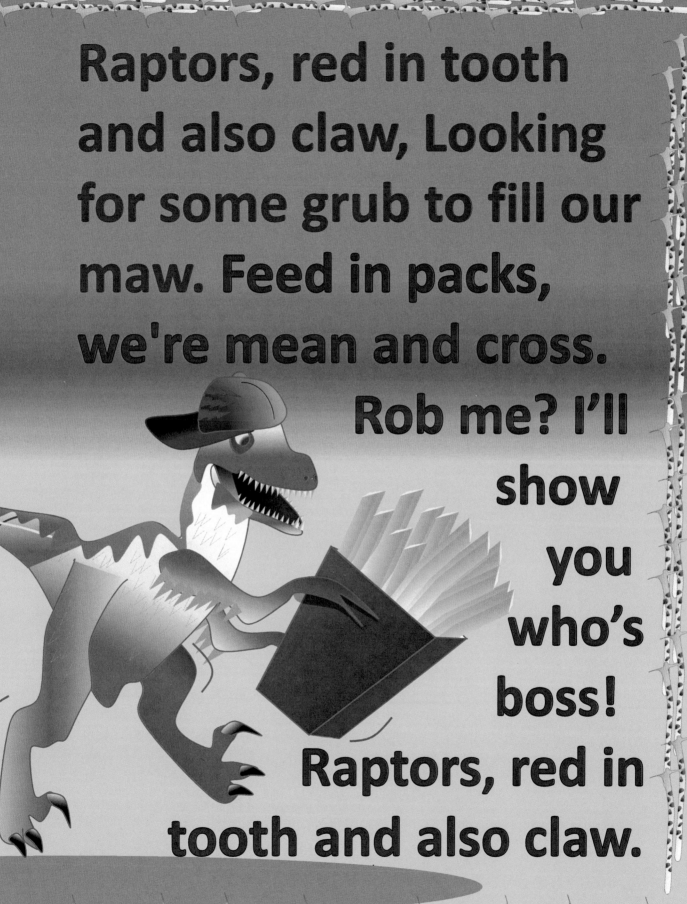

Raptors, red in tooth and also claw, Looking for some grub to fill our maw. Feed in packs, we're mean and cross. Rob me? I'll show you who's boss! Raptors, red in tooth and also claw.

Brontosaurus here, how do you do?
Leaves and grass I swallow, never chew.
   I rake the foliage with my teeth,
   And leave no tasty greens beneath.
      Brontosaurus here, how do you do?

Pachycephalosaurs,
with heads of bone,
Fight those who will not leave us alone.
Or bash our brothers on the crown—
Our eyes go
spinning
round and
round.
Pachycephalosaurs,
with heads of bone.

We Triceratops with gaudy frills—
Part protection, partly just for thrills.
Try to break our backs, we'll tip
Our crowns and poke you in the lip.
We Triceratops with gaudy frills!

Pterodactyls taking to the air,
Searching for a little snack to snare.
Soaring, swooping, spearing, stabbing,
Fishing, clamming, birding, crabbing,
Pterodactyls taking to the air!

What's an ancient Archaeopteryx?
Parts of bird and dino in a mix.
Beak of bird and plumes for flight—
It gave its prey a nasty fright!
That's an ancient Archaeopteryx.

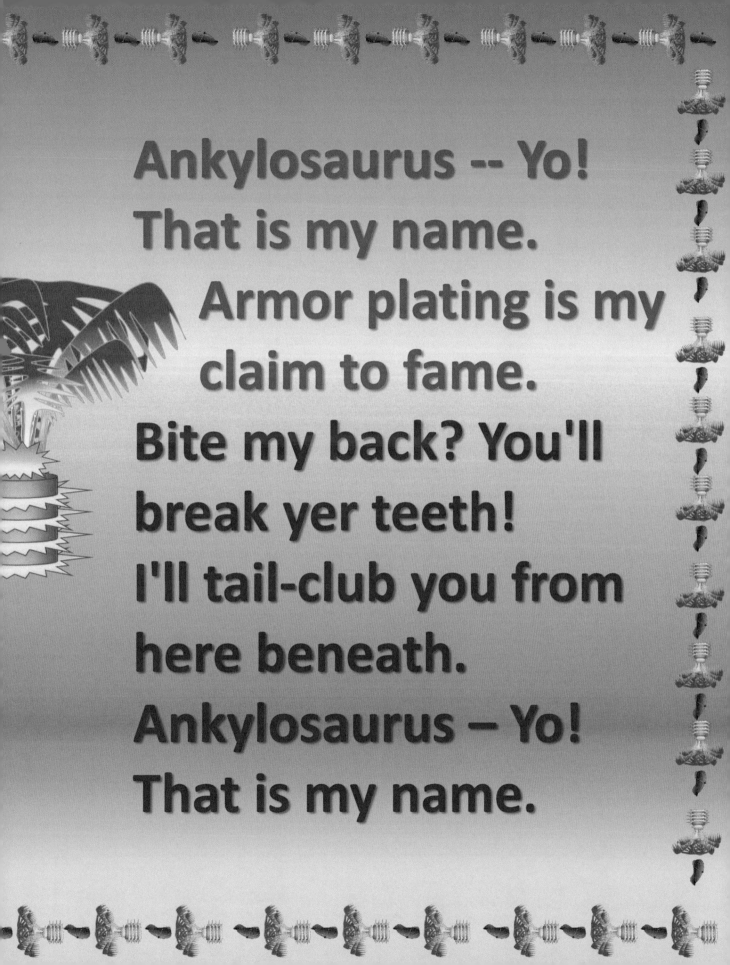

Ankylosaurus -- Yo!
That is my name.
Armor plating is my claim to fame.
Bite my back? You'll break yer teeth!
I'll tail-club you from here beneath.
Ankylosaurus – Yo!
That is my name.

Waiting for her clutch of eggs to hatch,
Duckbill mama watches out to catch
Robbers who her eggs would steal
To make themselves a tasty meal.
To a watchful mama they're no match!

Dinos conquered everywhere with ease.
Ichthyosaurs were masters of the seas.
Gobbled fishes by the score,
Their bellies full, they wanted more!
Ichthyosaurs were masters of the seas.

Spinosaurus sails, they win us praise
For their looks, and help us catch some rays.
Rockin' a Moroccan beach, you're
Sure to find no cooler creature.
Spinosaurus sails, they win us praise.

Dino looks? It's bound to end in tears:
Softer flesh like muscle disappears.
Wattles, combs and dorsal humps,
Earlobes, lips and fatty bumps...
Oh, to go back 80 million years!

Falling rocks from space are mostly small.
But *this* one—a gigantic flaming ball!
Fire, flood and heat it brought,
Global dark, as I was taught,
And swept away the dinos—nearly all.

Dinosaurs have mostly gone away—
Bones in great museums on display.
Yet Rex and Bronto still live on
In robin, gull and gentle swan.
Dinos live among us here today!

Notes

Though *I am a Tyrannosaurus Rex!* contains obvious elements from the present—knives, forks, wrestling shorts and crosswords to name a few—I was careful to include real dino info on every page.

Tyrannosaurus (Tie-RAN-uh-SORE-uss): Whether T. Rex was a fearsome predator, an opportunistic scavenger or some mix of the two, the sight of its powerful running legs and massive teeth would have unsettled the calmest of its fellow dinosaurs. Here, I show T. Rex in proper scavenging mode: eating something another animal had killed. Kind of like us, right?

Stegosaurus (STEGG-uh-SORE-uss): Stegosaurus really did have a small brain in comparison to other dinos. Maybe not quite walnut-sized, but about the size of a dog's—that, in a body that weighed 10,000 pounds! But then, it didn't have much to think about—just eating plants and swinging its tail if a meat-eater got too close. *Plants shown: Brachyphyllum (brown) and horsetail (green)*

Velociraptor (vuh-LOSS-uh-RAP-tor): One thing that will surprise fans of *Jurassic Park* is that the film's human-sized velociraptors were an enormous exaggeration. In reality, velociraptors might have been tall enough to bite an adult human on the bum—they were the size of a turkey, but that's about it. Another surprise: it's all but certain that all velociraptors were covered in feathers! Look carefully—I suggest that in my illustration. But pack-like behavior is unlikely, despite the outfits worn by our trio. It could be that many velociraptors converged at the scene of a kill to scavenge meat, but not because they coordinated their plans. Imagine them hopping around, tearing off chunks of flesh, squawking and fighting with each other—more like seagulls fighting over a dropped box of french fries than like wolves coordinating a sneak attack on an unsuspecting sheep.

Brontosaurus (BRONT-uh-SORE-us): Two debates shaped our understanding of bronto facts. First, its name. Bronto was discovered twice in the 1870s by rival scientists who seemed more interested in fighting each other than getting their facts straight. The scientist who first published his findings about the newly-found dinosaur called it "Apatosaurus," and that name (unbeknownst to the public) was its official name for decades. But the second scientist had the good fortune to work at the American Museum of Natural History. He labeled the dinosaur "Brontosaurus." And since many more people visit museums than read scientific journals, "Brontosaurus" was how the public came to know the dinosaur. "Brontosaurus" was thus burned into the imaginations of millions of kids. The name found its way into our culture via gas station logos, postage stamps and even cartoons like *The Flintstones,* where Fred Flintstone was a "bronto-crane" operator. The debate seemed settled until 2015, when further studies suggested that Apatosaurus and Brontosaurus were different species of dinosaur after all. I used "Brontosaurus" mostly for sentimental reasons, and in line with recent dino scholarship. But be prepared to have ferocious arguments with your favorite 6-year-old dino lover!!

The second debate about brontos is whether they could stand on their hind legs. This was not an issue when I was a kid in the 1960s. We all thought that brontosaurs were heavy and stupid and had to live semi-afloat in swamps to keep their weight off their legs. But soon after, those old ideas were challenged. Scientists suggested that brontos lived on dry land, stood on their hind legs to eat from high trees and actually *stomped* their predators! Today, the debate continues, and scientists are studying bronto skeletons to see if they held their necks upright or straight forward. I show brontos in two possible postures—reaching up by standing on hind legs, and swinging their long necks in an arc to graze on nearby vegetation.

Pachycephalosaurus (PACK-ee-SEPH-uh-luh-SORE-uss): The name means thick-headed lizard, and no wonder. It seems certain that these dinos butted domes with others, or at least threatened to. But whether it was their own species (as in my illustration) or in combat with other species is hard to pin down.

Triceratops (try-SARE-uh-tops): Surprise! Butterflies have existed for 140 to 200 million years—long enough in the past for a baby Triceratops to be lured away from her distracted mama!

Pterodactyl (TARE-uh-DACK-till): Their name literally meaning "winged finger," these creatures were not actually dinosaurs, but their distant cousins. Pterodactyls were small—with a wingspan of only three feet—but they were part of a larger family of pterosaurs (TARE-uh-sores), some of which had wingspans of 30 feet. That's as wide as a small airplane! Pterodactyl wings were made from stretched skin, but some pterosaurs were actually covered with short fur. Pterodactyls went extinct 136 million years ago, so some of the pterodactyl-like creatures on these pages are of their pterosaur cousins, which lived until all dinosaurs went extinct around 66 million years ago.

Archaeopteryx (arky-OPP-terr-icks): This creature, whose name means "old wing," was the first fossil of a winged and feathered dinosaur ever found. It seemed to bridge the gap between ancient dinosaurs and modern birds. Like dinosaurs, it had teeth and a bony tail; like modern birds, it had a wishbone and flight feathers. It wasn't a large bird—maybe the size of a crow. Our archaeopteryx is bearing down on a group of small mammals, our ancestors, one of whom is too intent on his berry-picking to notice danger swooping in from above.

Plants shown: Wollemi pine

Ankylosaurus (ankle-oh-SORE-uss): its name meaning "fused lizard" or "stiff lizard", for the likely awkward gait of this heavily armored animal, Ankylosaurus would have been hard to find in its time. It seems to have been rather rare, maybe because, with its impressive armor discouraging any hungry T. Rex, it didn't need to have huge numbers of descendants to keep its population stable. It seems to have been a slow mover, munching on whatever plants it came across. But it could explode in a moment of tail-swinging fury if a predator came unwisely close. I gave my beast a scally cap to suggest an easy-going personality that could erupt into action when provoked.

Plants show: ferns and Cycas (which still exists in India)!

Duckbill and Oviraptor (OH-vuh-RAP-ter): In my imagining of the scene, the egg-thief Oviraptor is trying its luck stealing eggs from a mama duckbilled dinosaur. It seems like easy pickings, but duckbills (a member of the large hadrosaur family) would not have succeeded as a species if they allowed every would-be robber to carry away their children! In Montana, a large collection of duckbill nests gives testimony to how duckbill adults and youngsters gathered together at nesting time—for mutual protection, most likely—in large nesting colonies like the one shown here. And a final note: Duckbills and oviraptors did live at the same time. My illustration is of the duckbill called Parasaurolopholus (PARE-uh-SORE-uh-LOFF-uh-luss) whose beautiful crest might have helped it recognize others of its own species.

Ichthyosaur (ICK-tee-oh-sore): Oops! Another ancient creature that is not a dinosaur, but a cousin adapted for the water. These dolphin-shaped reptiles ranged in size from three feet to monsters 48 feet long—as long as a city bus.

Spinosaurus (SPY-no-SORE-us): Did they use their sails to attract pretty or handsome spinosaurs? Did they use them to warm up on cool days? Or cool down on hot days? Those are some of the questions scientists still have about these beautiful creatures. They lived on water and land, so I show one cavorting in the waves. One thing we can be sure of: fossil evidence that they listened to the radio or enjoyed word searches has yet to be found!

The Bronto Bunch: The good news about dinosaurs is that we have found lots of their bones. But their softer bits don't fossilize (turn into rock) very well. So, it's a rare fossil that shows much beyond a skeleton. There's a fair amount of guesswork when trying to show how dinosaurs actually looked—their colors, internal organs and other soft bits that decay before being fossilized. And how about their habits—whether they lived alone or in groups, the noises they made, what they ate? Answering these questions requires scientists to make educated guesses—some of them very clever! Just to get you thinking, I started with a basic brontosaurus skull and added features cribbed from modern animals. If you can't guess what they are, here is the list:

Skull	Feathers, like a bird	Ears, like an elephant
Hump, like a camel		Inflatable throat sac, like a bullfrog
Wattles, like a turkey	Trunk, like an elephant and lips	Comb, like a rooster

The end of almost everything: One of the thrills of my lifetime was learning what actually happened to the dinosaurs—how they went extinct. When I was a kid, people thought that the dinosaurs were too slow and stupid to survive. We figured that with our smart and quick mammal ancestors around, the dopey dinosaurs didn't stand a chance. But really, before they suddenly disappeared, dinosaurs had been a hugely successful line of animals for *nearly 200 million years*! Our mammal ancestors had been happy just to survive in their enormous shadows.

Then, in 1980, an enormous crater was found in Central America, a crater that could only have been formed when an enormous asteroid—a space rock 6 to 9 miles across—collided with Earth. It was found that the asteroid punched a hole in the Earth's crust at *exactly* the same time—66 million years ago—that the dinosaurs disappeared. Scientists now agree that the simultaneous arrival of the asteroid and the departure of the dinosaurs was no coincidence.

They have calculated that the collision kicked up huge ocean waves, fiery rains of molten rock, forest fires and other changes that killed off many of the animals alive at that time. The dinosaurs, it turns out, were not stupid, slow or sickly. This wildly successful group of animals, which had conquered the land, air and seas around the Earth, had been all but exterminated by a global catastrophe that wiped out 75% of all species on the planet.

The illustration shows Earth just before the impact, viewing the Yucatan Peninsula of Central America from the north. A red circle just offshore shows the area that would soon become the Chicxulub (CHIX-ah-loob) crater – a 60-mile wide, 20-mile deep monster hole in the ground. The crater started flooding within 30 minutes of impact and remained hidden for the next 66 million years. Some of the rock blown out of the crater may have reached the moon. That's quite a blast!

Today: Among the survivors of the asteroid's impact (turtles, alligators and small mammals) were birds, which until that time were just another variety of dinosaur. That makes birds the only kind of dinosaur still alive today! In time, all of the animals that survived bounced back. Now, 66 million years after one of the worst disasters in our planet's history, there are 10,000 species of small flying dinosaurs...I mean, birds! That's an amazing comeback, and an exciting window into the past. For whenever you see a bird—whether it is hopping across a lawn or soaring above a beach or floating in a pond—you are seeing an animal whose ancestors might have made their living in the same way, alongside their gargantuan dinosaur cousins who ruled the Earth!

Easter Eggs

Easter Eggs are hidden treats scattered throughout a work of art. Here's a list of the eggs I hid in these pages.

All: Every page has a pair of brackets (]]) that shows the approximate height of an adult human and a 7-year-old child in comparison to the animal nearest the brackets. It's surprising how *small* were some of the dinos (pterodactyl, velociraptor and archaeopteryx) and how *enormous* were others (duckbill, bronto and spinosaurus)!

On the **Tyrannosaurus Rex** page, the stegosaurus in the rear is perplexed to see stego scales on T. Rex's plate. And for good reason: stegosaurus had been extinct for 20 million years before T. Rex came along. Tell that to your kid and watch them trumpet how "The book was wrong!" to all within earshot. There are many ways to learn!

The small brain of the main **stegosaurus** is visible.

The **raptors** are sporting Red Sox colors on their caps. Go Sox!

The middle **brontosaurus** is showing its gastroliths—literally stomach-stones. Bronto had no time to chew its food. It just gobbled tons of plants and branches and let stomach juices and the gastroliths grind the materials into a digestible pulp. Chickens do the same today, though of course with much smaller stones.

On the page showing the **asteroid** approaching Earth, there's a red circle where the asteroid struck, just off the coast of the Yucatan Peninsula, representing the 60-mile width of the resulting crater.

Bibliography

Here's a list of books that formed my ideas about the dinosaurs, some of which found their way into *this* book!

Bully for Brontosaurus: Reflections in Natural History by Stephen Jay Gould, published in 1991 by W. W. Norton and Company. The cover shows a hilariously outdated (intentionally so) depiction of slow, brutish brontosaurus, languishing in a swamp because of its otherwise "unsupportable" bulk. Gould covers the Bronto-Apatosaurus debate in lip-smacking detail.

The Dinosaur Heresies: New Theories Unlocking the Mystery of the Dinosaurs and Their Extinction by Robert T. Bakker, PhD, published in 1996 by Zebra Books. Bakker seems intent on demolishing every sacrosanct bit of "knowledge" about dinos. He argues that dinos were warm-blooded, that birds evolved from dinosaurs and that dinosaurs were not sluggish layabouts.

The Rise and Fall of the Dinosaurs: A New History of Their Lost World by Steve Brusatte, published in 2018 by William Morrow. Brusatte is at his best explaining how the great eras of Earth's history came to be. Short answer? Global cataclysms—oxygenation, volcanic eruptions, asteroids, etc.—ended each period, wiping out enormous numbers of species and clearing the way for the next crop.

Text

The seeds for this book were sown back in 1992, in a little ditty I sang to my two sons, John and James, at bedtime. That incarnation of *I am a Tyrannosaurus Rex!* is identical to the first bit of verse in this book. In later years I added a another verse or two. But it wasn't until I considered turning the song into a book in 2018 that things really took off, seeing me complete the rest of the verses, usually as I created the illustrations.

The Illustrations

As with my previous book, *Bernie and the Day the Icebergs Melted*, I used Microsoft PowerPoint for the illustrations. I relied less this time on pre-formed shapes (circles, squares and cylinders) and more on PowerPoint's freeform shape tool and its amazing ability to stretch and twist shapes. As before, color gradients—shadings from one hue to another—found their way into many illustrations.

About the Author

When he is not cooking up new books, Jean Pouliot works as an IT product owner for a large insurance company. He contributes stories, songs, essays and information about his writing projects to his web site, www.jeanedouardpouliot.com. Go there to learn more about how this book was created!

Jean grew up speaking Canadian French in a French-Catholic section of Manchester, NH. He and Carol, his wonderful wife of 35+ years, live in Massachusetts. John and James, their two terrific kids, are making their own marks on the worlds of political advertising and news writing.

Jean has a BA in Psychology from the University of New Hampshire and a Masters in Catholic Studies from St. John's Seminary in Brighton, MA. His interests include hiking, photography, movies and plays, US history and everything scientific: cognition, paleontology, astronomy and evolutionary biology. He has traveled all over the US and Eastern Canada as well as Rome, London, Paris, Normandy, South Africa and the Netherlands.

Also by Jean Pouliot:
Bernie and the Day the Icebergs Melted

Made in the USA
Middletown, DE
27 September 2023